Experiments with LIQUIDS

Christine Taylor-Butler

Heinemann
LIBRARY
Chicago, Illinois

www.heinemannraintree.com
Visit our website to find out
more information about
Heinemann-Raintree books.

To order:
☎ Phone 888-454-2279
💻 Visit www.heinemannraintree.com
to browse our catalog and order online.

Edited by Rebecca Rissman, Dan Nunn, and Catherine Veitch
Designed by Richard Parker
Picture research by Tracy Cummins
Originated by Capstone Global Library
Printed and bound in China by South China Printing Co. Ltd

15 14 13 12 11
10 9 8 7 6 5 4 3 2 1

Library of Congress Cataloging-in-Publication Data
Taylor-Butler, Christine.
  Experiments with liquids / Christine Taylor-Butler.—1st ed.
     p. cm.—(My science investigations)
  Includes bibliographical references and index.
  ISBN 978-1-4329-5361-4 (hc)—ISBN 978-1-4329-5367-6
(pb) 1. Liquids—Experiments—Juvenile literature. 2. Science
projects—Juvenile literature. I. Title. II. Series.

  QC145.24.T39 2012
  530.4'2—dc22   2010042652

Acknowledgments
We would like to thank the following for permission to
reproduce photographs: Heinemann Raintree p. 9, 10, 11,
12, 13, 14, 16, 17, 18, 19, 20, 21, 24, 25, 26 (Karon Dubke);
Shutterstock p. 4 (© Constantine Androsoff), 5
(© niderlander), 6 (© Goodluz), 23 (© Luis Santos), 28
(© cubephoto), 29 (© Laurence Gough).

Cover photograph of a girl looking at a vial of liquid
reproduced with permission of Photolibrary (Hill Street
Studios). Background photograph of bubbles reproduced with
permission of Shutterstock (Rich Carey).

Special thanks to Suzy Gazlay for her invaluable help in the
preparation of this book. We would also like to thank Ashley
Wolinski for her help in the preparation of this book.

Contents

Different Liquids 4

How Scientists Work 6

Changing States of Matter 8

Liquid Rainbows 12

Making Solutions 16

Can It Float? 20

Soak It Up! 24

Your Turn! 28

Glossary 30

Find Out More 31

Index 32

Some words are printed in bold, **like this.**
You can find out what they mean by looking
in the Glossary.

Different Liquids

Liquids are all around us. Milk, paint, and the ink in a pen are liquids. So are the oil and gas in a car.

The most common liquid is water. It is found in oceans, lakes, rivers, and streams. Water is also found in animals, plants, and other living things. Scientists who study water are called **hydrologists**.

How Scientists Work

Scientists start with a question about something they **observe**, or notice. They gather information and think about it. Then they make a guess, or **hypothesis**, about a likely answer to their question. Next they set up an **experiment** to test their hypothesis. They look at the **data**, or **results**, and make a decision, or **conclusion**, about whether their hypothesis is right or wrong.

Whether their hypothesis is right or wrong, scientists still learn from each experiment.

How To Do an Experiment

1. Start a **log**. Write down your **observations**, question, and hypothesis.
2. Plan step-by-step how you can test the hypothesis. This is called the **procedure**.
3. Carry out the experiment. **Record** everything that happens. These are your observations.
4. Compare your results with your hypothesis. Was your hypothesis right or wrong? What did you learn? The answer is your conclusion.

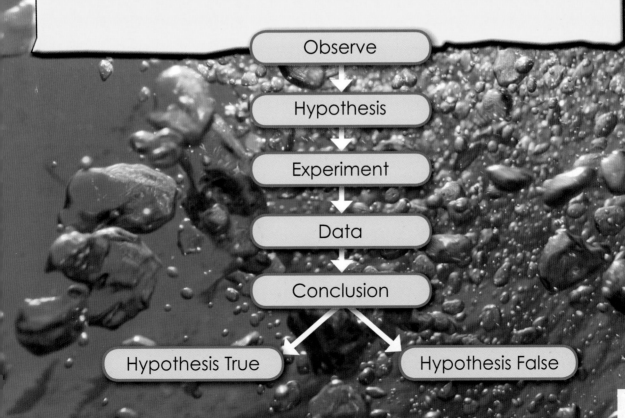

Observe

↓

Hypothesis

↓

Experiment

↓

Data

↓

Conclusion

Hypothesis True — Hypothesis False

Changing States of Matter

Everything is made up of **matter**. Matter comes in three forms. Solids take up a certain amount of space and have their own shape. Liquids take up a certain amount of space, but they can take the shape of whatever container they are in. Gases do not have a shape and can take up any amount of space.

ice (solid)

water (liquid)

water (gas)

Water is the only kind of matter that is found naturally in all three states, or forms, on our planet.

Temperature can make matter change to a different form, or state. Heat can turn a solid into a liquid, or a liquid into a gas. Cooling can turn a gas into a liquid, or a liquid into a solid. For some kinds of matter, the temperatures need to be very high or very low for changes to happen.

Let's see how this works with water.

You'll need these things for this **experiment**.

Procedure

1. Ask an adult to help pour hot water from a faucet into a plastic bottle until it is half full.
2. Replace the cap and shake the bottle well.
3. Remove the cap. Cover the opening with an ice cube. What happens inside the bottle?
4. Replace the cap. Put the bottle and a thermometer in a freezer.
5. Check the bottle every hour. What is the temperature? How long does it take the water to freeze? **Record** your **observations**.

Water (Gas)

Water (Liquid)

6. Place the bottle and thermometer on a table. At what temperature does the ice melt and become a liquid again? Write your answers in your **log**.

The Science Explained

Some of the hot water turns into water **vapor** and rises. When it touches the ice cube, it cools and changes back into water. The water in the freezer freezes and turns into ice.

Liquid Rainbows

Not all liquids are the same. Liquids have **properties**, or characteristics, that cause them to behave differently. Some liquids are more **dense** than others because they contain more **matter**.

Hypothesis

Some liquids will form layers rather than mixing with other liquids.

You will need these things for this **experiment**.

Procedure

1. Put about 1 inch of corn syrup into a glass. Add 3 drops of blue food color. Mix well.
2. Put about 1 inch of water into a second glass. Add 3 drops of red food color. Mix well.
3. Slowly add the red water to the blue syrup. **Record** your **observations**.

4. Put about 1 inch of vegetable oil into a third glass. Add 3 drops of one food color to the oil. Mix well. What happens?

5. Slowly add the vegetable oil to the glass containing water and corn syrup. Wait and watch what happens. Draw a picture of your **results**.

6. Did all of the liquids mix? **Record** your **observations** in your **log**.

Be careful when using food coloring. It stains!

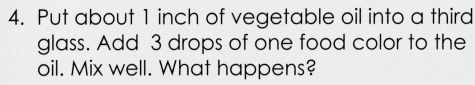

The Science Explained

The heaviest (most **dense**) liquid will sink to the bottom of the glass. The lightest (least dense) will form a layer on top.

What I did	What I observed
Added water to syrup	
Added vegetable oil	

Record your results in a chart like this in your own log.

Making Solutions

When some solids are stirred into a liquid, they **dissolve**, or break into pieces that are too small to see. This mixture is called a **solution**. As more of the solid is added, it will stop dissolving and begin to pile up on the bottom. This happens when the solution cannot hold any more of the tiny pieces.

Hypothesis

Sugar will dissolve more easily than salt, pepper, or cinnamon in water.

You will need these things for this **experiment**.

Procedure

1. Measure one cup of warm water and pour it into a glass.

2. Add one tablespoon of sugar. Stir 20 times. What happens?

3. Repeat steps 1 and 2 three more times, using salt, pepper, and cinnamon. **Record** your **observations.** Which solids dissolved most easily?

Which dissolves more easily, sugar or salt?

4. Stir more sugar into the sugar **solution**, one tablespoon at a time. How much can you add before it stops **dissolving** and settles on the bottom of the glass?

5. Try stirring more salt into your salt solution. How many tablespoons does it take to get the same **results**?

The Science Explained

The **properties**, or characteristics, of salt and sugar allow them to dissolve into water. The properties of pepper and cinnamon do not.

Sugar that has not dissolved

Wow!

1. Fill a glass halfway with water. Slide a raw egg into the glass. Does it float or sink?
2. Remove the egg. Stir in two tablespoons of salt. Put the egg back. What happens?
3. Slowly add more water. What happens now?

The Science Explained

Adding salt to the water gradually makes the solution more **dense** until the egg floats. Adding water makes it less dense again, and the egg sinks.

If you don't see a change during step 2, add more salt until you see something happen.

Can It Float?

Some objects float on water. Others sink to the bottom. If an object is less **dense** than water, it will float. If it is more dense, it will sink.

Can you guess which objects will float?

Find a variety of objects to test, such as coins, pencils, and paperclips.

Procedure

1. Hold each object in your hand, one at a time. Guess whether it will sink or float.

2. List each object in your **log**. **Record** your guess.

3. Fill a bowl halfway with water. Place each object on the surface of the water. Does it sink or float? Record your **results**.

Record how many times your guesses were correct.

4. Choose more objects. Guess whether you think they will float. **Record** your findings. Did any objects surprise you?

Object	Predict: sink or float?	Observe: sank or floated?
~~	~~~~~	~~~~~
~~	~~~~~	~~~~~

Make a chart like this to record your results in your **log**.

Try this!
Hypothesis
Fresh fruit floats.

Procedure
1. Find several different pieces of fresh fruit. Test each one to see if it floats.
2. Record your **observations**.
3. Check your **results**. Was your **hypothesis** right or wrong?

Guess whether or not the fruit will float first, before testing it. Were you correct?

23

Soak It Up!

When you use a paper towel to clean up a spill, the paper towel **absorbs** the spill. It is pulled up from the surface into the paper towel.

Hypothesis

A paper towel can absorb water from one container and move it to another.

You will need these things for your **experiment**.

Procedure

1. Fill a glass with water, leaving some space at the top.

2. Twist a paper towel to make a thin tube. Fold the tube in half.

3. Put one end into the glass of water. Put the other end into an empty measuring cup. **Record** your **observations**.

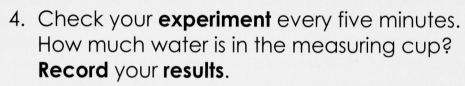

4. Check your **experiment** every five minutes. How much water is in the measuring cup? **Record** your **results**.

5. Leave the experiment overnight. How high is the water in the glass compared to the water in the measuring cup?

Record your results each time you check on your experiment.

The Science Explained

Water is **absorbed** through small spaces in the paper towel. When the water reaches the top of the glass, **gravity** pulls it down through the paper towel into the measuring cup. The water stops moving when the water levels in the glass and the measuring cup are the same.

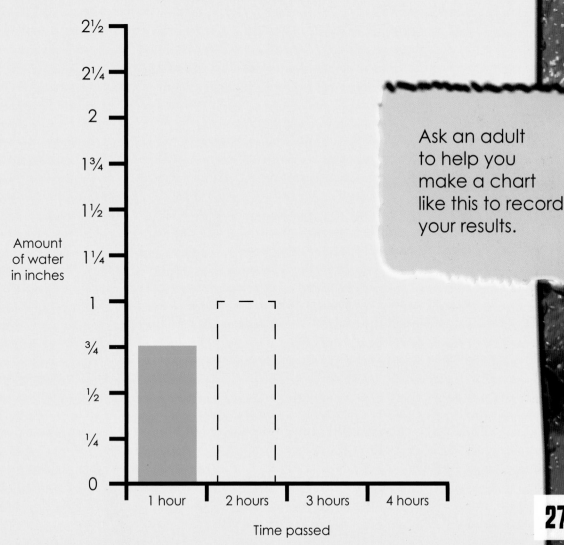

Amount of water in inches

2½
2¼
2
1¾
1½
1¼
1
¾
½
¼
0

1 hour   2 hours   3 hours   4 hours

Time passed

Ask an adult to help you make a chart like this to record your results.

Your Turn!

Hydrologists work all around the world. They study water in forests, on farms, and in oceans. They help to design dams, reservoirs, and water systems for cities. Hydrologists find ways to keep water clean and healthy.

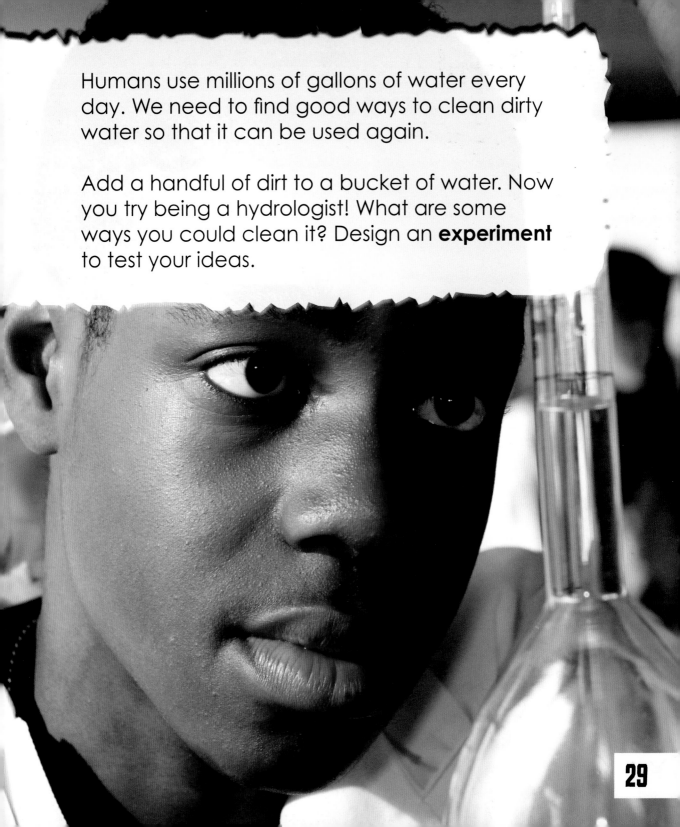

Humans use millions of gallons of water every day. We need to find good ways to clean dirty water so that it can be used again.

Add a handful of dirt to a bucket of water. Now you try being a hydrologist! What are some ways you could clean it? Design an **experiment** to test your ideas.

Glossary

absorb suck up, or take in

conclusion what you learn from the results of an experiment

data information gathered in an experiment

dense contains a lot of matter

dissolve break down into tiny pieces

experiment organized way of testing an idea

gravity force that pulls things toward Earth

hydrologist scientist who studies water

hypothesis suggested statement or explanation that can be tested

log written notes about an experiment

matter anything that takes up space

observe watch, or notice something

observation something you notice, or observe, with any of your five senses

procedure steps followed to carry out an experiment

properties traits or characteristics that can be observed and measured

record draw or write something down

results what happens in an experiment

solution substance formed when a solid dissolves in a liquid

vapor gas form of water, which is steam

Find Out More

Books

Adamson, Thomas K. and Heather. *How Do You Measure Liquids*. Mankato, Minn.: Capstone, 2010.

Boothroyd, Jennifer. *What Floats? What Sinks? A Look at Density*. Minneapolis, Minn.: Lerner, 2010.

Royston, Angela. *Solids, Liquids and Gasses*. Chicago: Heinemann, 2008.

Websites

American Museum of Natural History
www.amnh.org/ology/?channel=water

NASA For Kids
kids.earth.nasa.gov/droplet.html

Wonderville—Discover the fun of science
www.wonderville.ca/browse/activities

Index

absorbing 24–27

cleaning dirty water 29

conclusions 6, 7

cooling 9

data 6, 7

dense liquids 12, 15

dense objects 20–23

dense solutions 19

dissolving 16, 17, 18

eggs 19

experiments 6–7

floating 20–23

fruit 23

gases 8, 9, 10

gravity 27

heat 9

hydrologists 5, 28

hypothesis 6, 7

ice 8, 11

layers, forming 12–15

liquid form of matter 8

log 7

matter 8–11, 12

mixing liquids 12–15

observations 6, 7

paper towels 24–27

procedure 7

properties 12–15, 18

recording 7

results 6, 7

salt 16, 17, 18, 19

scientists 5, 6, 28

sinking 20, 21

solids 8, 9, 16

solutions 16–19

states of matter, changing 8–11

sugar 16–19

temperature 9, 10

water 5, 8, 10–11, 13–14, 16–19, 20–27

water vapor 11